FIBONACCI-LIKE
SEQUENCES

Fibonacci-Like Sequences

Sequences

A Scientific Approach

Edgar M. Alexander

Library of Congress Control Number:		2011918974
ISBN:	Hardcover	978-1-4653-8541-3
	Softcover	978-1-4653-8540-6
	Ebook	978-1-4653-8542-0

To order additional copies of this book, contact:
Xlibris Corporation
1-888-795-4274
www.Xlibris.com
Orders@Xlibris.com
106592

EDGAR M. ALEXANDER is a native of Tuskegee, Alabama, and that is where he spent his early childhood days. Even at his early age, he was fascinated by numbers. This was primarily due to the idea that mathematics always had an answer, and the answer was universal, where everyone could agree. In 1964 he left home to become part of the first ABC program (A Better Chance) at Dartmouth College.

Having been accepted to attend private school in the fall of that year, he started classes at Wilbraham Academy. Here he finished his high school requirements.

At Tufts University, he majored in mathematics and minored in economics. This is where he was first introduced to calculus. Mathematics continued to be his major interest.

Upon graduating from Tufts, he continued his education with the following: master's degree from Howard University in urban systems engineering and several courses in operations research in a doctorate program at George Washington University. That concludes his education background. Since then he has worked for thirty-five years in the United States government.

In addition to this paper, he recently wrote *In Search of Pi*.

That book is a new creation that develops the circle as a function. Both of these documents are related and are recommended to be viewed together.

The main purpose of this paper is to introduce the reader to an alternative to proof by induction. While this document is not meant to replace existing material on the subject, it does offer new, innovative concepts. My document will introduce the reader to a method of defining all Fibonacci numbers with a single parameter. The result of this allows formula to be proven directly. Proofs are therefore easily understood. As a classroom tool for teaching Fibonacci numbers and Lucas numbers, this paper is a necessity.

Beyond that, the same principles are extended to the universe of all Fibonacci-like sequences. The reader will be able to explore relationships of other similar sequences.

Two formulas used in the development of this paper were actually created and copyrighted in another of my manuscripts, *In Search of Pi*. When given F(n), F(n + 1) can be computed exactly and directly as follows:

$$F_{2n} = (F_{2n-1} + (5 * F^2_{2n-1} - 4)^{\frac{1}{2}}) / 2$$
and
$$F_{2n+1} = (F_{2n} + (5 * F^2_{2n} + 4)^{\frac{1}{2}}) / 2$$

Fibonacci numbers and Lucas numbers have been shown to have many common characteristics. However, there are an infinite number of such sequences.

I will now establish a description of all such sequences and a scientific approach for investigating these sequences.

A Scientific Approach to Fibonacci-Like Numbers

EDGAR M. ALEXANDER

I will first give my definition of Fibonacci-like numbers:

A sequence of numbers $a_n, a_{n+1}, a_{n+2},$ where $a_n + a_{n+1} = a_{n+2}.$

All these sequences have a direct relationship with the golden ratio, $(1 + 5^{.5})/2 = 1.618034.$

Observation

Our observation of the characteristics of these sequences is made through the following:

I. Fibonacci numbers

   a) The sequence is

$$F_1 = 1,$$
$$F_2 = 1,$$
$$F_3 = 2,$$
$$F_4 = 3,$$
$$F_5 = 5,$$
$$F_6 = 8.$$

   b) The ratio of consecutive members is
       $F_n, F_{n+1}/F_n$

   1) 1,
   2) 1, $F_2/F_1 = 1,$
   3) 2, $F_3/F_2 = 2,$
   4) 3, $F_4/F_3 = 1.5,$

5)   5, $F_5 / F_4 = 1.666666\ldots$,
6)   8, $F_6 / F_5 = 1.6$

. . .

. . .

. . .

19) 4181,
20) 6765, $F_{20} / F_{19} = 1.618034$.

We can observe the following:

F_{2n+1} / F_{2n} is always greater than the golden ratio.
F_{2n} / F_{2n-1} is always less than the golden ratio.

The limit of the ratio F_k / F_{k-1} is always the golden ratio or $1.618034\ldots$

C) We note that there are formulas for computing Fibonacci numbers directly:

$$F_{2n} = (F_{2n-1} + (5 * F^2_{2n-1} - 4)^{\frac{1}{2}}) / 2$$
and
$$F_{2n+1} = (F_{2n} + (5 * F^2_{2n} + 4)^{\frac{1}{2}}) / 2$$

can be useful in this effort.

F_k
1
1, $F_2 = (1 + (5 * 1^2 - 4)^{\frac{1}{2}}) / 2 = (1 + 1) / 2 = 1$
2, $F_3 = (1 + (5 * 1^2 + 4)^{\frac{1}{2}}) / 2 = (1 + 3) / 2 = 2$
3, $F_4 = (2 + (5 * 2^2 - 4)^{\frac{1}{2}}) / 2 = (2 + 4) / 2 = 3$
5, $F_5 = (3 + (5 * 3^2 + 4)^{\frac{1}{2}}) / 2 = (3 + 7) / 2 = 5$

EDGAR M. ALEXANDER

$$8, F_6 = (5 + (5 * 5^2 - 4)^{1/2})/2 = (5 + 11)/2 = 8$$

Here is an illustration of the ratio of consecutive terms:

$F_k , F_{n+1}/F_n$

1

$1, F_2/F_1 = (1 + (5 - 4/F_1^2)^{1/2})/2 = (1 + 1)/2 = 1 < \varphi$

$2, F_3/F_2 = (1 + (5 + 4/F_2^2)^{1/2})/2 = (1 + 3)/2 = 2 > \varphi$

$3, F_4/F_3 = (1 + (5 - 4/F_3^2)^{1/2})/2 = (1 + 2)/2 = 1.5 < \varphi$

$5, F_5/F_4 = (1 + (5 + 4/F_4^2)^{1/2})/2 = 1.6666 \ldots > \varphi$

$8, F_6/F_5 = (1 + (5 - 4/F_5^2)^{1/2})/2 = 1.6 < \varphi$

$13, F_7/F_6 = (1 + (5 + 4/F_6^2)^{1/2})/2 = 1.625 > \varphi$

We observe that if the leading term is even, then the ratio is less than φ, 1.618034. And if the leading term is odd, then the ratio is greater than φ. However, with Lucas numbers we observe that if the leading term is even, then the ratio is greater than 1.618034 ... And if the leading term is odd, then the ratio is less than φ. Also, the limit of consecutive ratios is the golden ratio.

II. Lucas numbers

a) The sequence is

$L_1 = 1,$

$L_2 = 3 = (L_1 + (5 * L_1^2 + 20)^{.5})/2 = (1 + (5 * 1 + 20)^{.5})/2 = 6/2,$

$L_3 = 4 = (L_2 + (5 * L_2^2 - 20)^{.5})/2 = (3 + (5 * 9 - 20)^{.5})/2 = 8/2,$

$L_4 = 7 = (L_3 + (5 * L_3^2 + 20)^{.5})/2 = (4 + (5 * 16 + 20)^{.5})/2 = 14/2,$

$L_5 = 11 = (L_4 + (5 * L_4^2 - 20)^{.5})/2 = (7 + (5 * 49 - 20)^{.5})/2 = 22/2,$

$L_6 = 18 = (L_5 + (5 * L_5^2 + 20)^{.5})/2 = (11 + (605 + 20)^{.5})/2 = 36/2.$

b) The ratio of consecutive terms yields the following results:

$$L_2/L_1 = (1 + (5 + 20/L_1{}^2)^{1/2})/2 > \varphi$$
$$L_3/L_2 = (1 + (5 - 20/L_2{}^2)^{1/2})/2 < \varphi$$
$$L_4/L_3 = (1 + (5 + 20/L_3{}^2)^{1/2})/2 > \varphi$$
$$L_5/L_4 = (1 + (5 - 20/L_4{}^2)^{1/2})/2 < \varphi$$

From our observations, we can now make a hypothesis that covers all Fibonacci-like sequences.

Hypothesis

1) Fibonacci-like series are composed of odd and even members.
2) The relationship of consecutive numbers (n_1, n_2, n_3) is $n_1 + n_2 = n_3$.
3) The limit of n_k/n_{k-1} as k approaches infinity is the golden ratio, $(1 + 5^{1/2})/2$.
4) Formulas can be created to directly compute n_{k+1} from n_k.

Proof
Here is an informal proof of the above hypothesis:

Fibonacci-Like Sequences
Define this sequence as having the following properties:

$$a_n, a_{n+1}, a_{n+2}, \text{ where } a_n + a_{n+1} = a_{n+2}$$

Then the following are true:

(1) $a_{n+1} = (a_n + (5 * a_n{}^2 + c)^{1/2})/2$ and

(2) $a_{n+2} = (a_{n+1} + (5 * a_{n+1}{}^2 - c)^{1/2})/2$, where c is constant.

Proof:

Let
$$c = (2 * a_{n+1} - a_n)^2 - 5 * a_n^2$$

Then

$$a_{n+1} = (a_n + (5 * a_n^2 + c)^{\frac{1}{2}}) / 2 = (a_n + ((2 * a_{n+1} - a_n)^2)^{\frac{1}{2}}) / 2 = 2 * a_{n+1} / 2 = a_{n+1}$$

This proves that c does create a_{n+1} from a_n.

Now show that c is the same for all odd subscripts of a. And c would have the negative value of that for all even subscripts:

Let $K_1 = c = (2 * a_{n+1} - a_n)^2 - 5 * a_n^2 = (a_{n+1}^2 - a_{n+1} * a_n - a_n^2) * 4$
Let $K_2 = (2 * a_{n+2} - a_{n+1})^2 - 5 * a_{n+1}^2 = (a_{n+2}^2 - a_{n+2} * a_{n+1} - a_{n+1}^2) * 4$

Then

$$(K_1 + K_2) / 4 = a_{n+1}^2 + a_{n+2}^2 - (a_{n+1} * a_n + a_{n+2} * a_{n+1}) - a_n^2 - a_{n+1}^2 = a_{n+2}^2 - (a_{n+1} *$$
$$a_n + a_{n+2} * a_{n+1}) - a_n^2 =$$
$$(a_{n+1} + a_n)^2 - (a_{n+1} * a_n + (a_{n+1} + a_n) * a_{n+1}) - a_n^2 =$$
$$a_{n+1}^2 + 2 * a_{n+1} * a_n + a_n^2 - a_{n+1} * a_n - a_{n+1}^2 - a_{n+1} * a_n - a_n^2 = 0$$
$$K_2 = -K_1$$
$$K_3 = -K_2 = K_1 \text{ or } K_3 = K_1$$
$$K_4 = -K_3 = -(-K_2) = K_2 \text{ or } k_4 = K_2$$
$$K_{2n} = K_2 \text{ for all } n \text{ and }$$
$$K_{2n+1} = K_1 \text{ for all } n. \text{ Therefore, } K_{2n+1}, K_{2n} \text{ are constant.}$$

For the second portion of this proof, show that the negative value of c will produce a_{n+2} from a_{n+1}.

Let $M = (a_{n+1} + (5 * a_{n+1}^2 - c)^{\frac{1}{2}}) / 2$ and let $c = (2 * a_{n+1} - a_n)^2 - 5 * a_n^2$.

Then $M = (a_{n+1} + (5 * a_{n+1}^2 - c)^{\frac{1}{2}}) / 2 =$

$(a_{n+1} + (5 * a_{n+1}^2 - (2 * a_{n+1} - a_n)^2 + 5 * a_n^2)^{\frac{1}{2}}) / 2 =$

$(a_{n+1} + (a_{n+1}^2 + 4 * a_{n+1} * a_n + 4 * a_n^2)^{\frac{1}{2}}) / 2 =$

$(a_{n+1} + ((a_{n+1} + 2 * a_n)^2)^{\frac{1}{2}}) / 2 =$

$(a_{n+1} + a_{n+1} + 2 * a_n) / 2 = a_n + a_{n+1} = a_{n+2}$ by definition.

$M = a_{n+2}$

Here, c is a constant and may be computed at any even member within the sequence. Finally we can easily show that the limit of the ratio of consecutive Fibonacci-like numbers is the golden ratio.

1) Let G_1, G_2, G_3 represent consecutive Fibonacci-like numbers.
2) Then $G_2 = (G_1 + (5 * G_1^2 + c)^{\frac{1}{2}}) / 2$ and $G_3 = (G_2 + (5 * G_2^2 - c)^{\frac{1}{2}}) / 2$
3) $G_2 / G_1 = (G_1 / G_1 + (5 * G_1^2 / G_1^2 + c / G_1^2)^{\frac{1}{2}}) / 2 =$
   $(1 + (5 * 1 + c / G_1^2)^{\frac{1}{2}}) / 2$.
4) The limit of c / G_1^2 as G_1^2 approaches infinity is 0.
5) The limit of G_2 / G_1 as G_1^2 approaches infinity is $(1 + 5^{\frac{1}{2}}) / 2$.
6) $G_3 / G_2 = (G_2 / G_2 + (5 * G_2^2 / G_2^2 - c / G_2^2)^{\frac{1}{2}}) / 2 =$
   $(1 + (5 * 1 - c / G_2^2)^{\frac{1}{2}}) / 2$.
7) The limit of c / G_2^2 as G_2^2 approaches infinity is 0.
8) The limit of G_3 / G_2 as G_2^2 approaches infinity is $(1 + 5^{\frac{1}{2}}) / 2$.

Conclusion

Note that the Fibonacci-like sequences have the following characteristics:

1) Where $c \neq 0$, the limit of the ratio of consecutive members is the golden ratio.
2) The sequence is made of even and odd members.
3) Where $c \neq 0$, the golden ratio is always bounded by the ratios of the two member types, (J_{2n}/J_{2n-1}) and (J_{2n+1}/J_{2n}).
4) The next member in the sequence may be computed directly through usage of the above proofs. Also, you can see that both Lucas and Fibonacci numbers are members of this sequence type. However, the number of this sequence type is infinite.

Examples

Here is an illustration of how to compute c for the following sequences:

1) $1, 1, 2, 3, 5 \ldots$ Fibonacci series

$$F_1 = 1, F_2 = 1,$$
$$c = (2 * a_{n+1} - a_n)^2 - 5 * a_n^2$$
$$c = (2 * 1 - 1)^2 - 5 * 1^2 = 1 - 5 = -4$$
$$F_{2n} = (F_{2n-1} + (5 * F_{2n-1}^2 - 4)^{\frac{1}{2}}) / 2$$
$$F_{2n+1} = (F_{2n} + (5 * F_{2n}^2 + 4)^{\frac{1}{2}}) / 2$$

2) $1, 3, 4, 7, 11 \ldots$ Lucas numbers
$$L_1 = 1, L_2 = 3,$$
$$c = (2 * a_{n+1} - a_n)^2 - 5 * a_n^2$$
$$c = (2 * 3 - 1)^2 - 5 * 1^2 = 25 - 5 = 20$$
$$L_{2n} = (L_{2n-1} + (5 * L_{2n-1}^2 + 20)^{\frac{1}{2}}) / 2$$
$$L_{2n+1} = (L_{2n} + (5 * L_{2n}^2 - 20)^{\frac{1}{2}}) / 2$$

$$* \quad * \quad *$$

3) 1, 5, 6, 11, 17 . . . test numbers

$T_1 = 1, T_2 = 5,$

$c = (2 * a_{n+1} - a_n)^2 - 5 * a_n^2$

$c = (2 * 5 - 1)^2 - 5 * 1^2 = 81 - 5 = 76$

$T_{2n} = (T_{2n-1} + (5 * T_{2n-1}^2 + 76)^{1/2}) / 2$

$T_{2n+1} = (T_{2n} + (5 * T_{2n}^2 - 76)^{1/2}) / 2$

As seen from the above examples, the formula for the next sequence member and the rate of divergence depend on the value of c. The following table gives the c values for a number of positive integer sequences.

1) 1, 4, 5, 9, . . . , $c = 44,$ $S_{2n} = (S_{2n-1} + (5 * S_{2n-1}^2 + 44)^{1/2}) / 2$

2) 1, 5, 6, 11, . . . , $c = 76,$ $S_{2n} = (S_{2n-1} + (5 * S_{2n-1}^2 + 76)^{1/2}) / 2$

3) 1, 6, 7, 13, . . . , $c = 116,$ $S_{2n} = (S_{2n-1} + (5 * S_{2n-1}^2 + 116)^{1/2}) / 2$

4) 1, 7, 8, 15, . . . , $c = 164,$ $S_{2n} = (S_{2n-1} + (5 * S_{2n-1}^2 + 164)^{1/2}) / 2$

5) 1, 8, 9, 17, . . . , $c = 220,$ $S_{2n} = (S_{2n-1} + (5 * S_{2n-1}^2 + 220)^{1/2}) / 2$

6) 1, 9, 10, 19, . . . , $c = 284,$ $S_{2n} = (S_{2n-1} + (5 * S_{2n-1}^2 + 284)^{1/2}) / 2$

7) 1, 10, 11, 21, $c = 356,$ $S_{2n} = (S_{2n-1} + (5 * S_{2n-1}^2 + 356)^{1/2}) / 2$

8) 1, 11, 12, 23, $c = 436,$ $S_{2n} = (S_{2n-1} + (5 * S_{2n-1}^2 + 436)^{1/2}) / 2$

values for the minus of the same integer sequences.

1) −1, −4, −5, −9, . . . , $c = 44,$ $S_{2n} = (S_{2n-1} - (5 * S_{2n-1}^2 + 44)^{1/2}) / 2$

2) −1, −5, −6, −11, . . . , $c = 76,$ $S_{2n} = (S_{2n-1} - (5 * S_{2n-1}^2 + 76)^{1/2}) / 2$

3) −1, −6, −7, . . . , $c = 116,$ $S_{2n} = (S_{2n-1} - (5 * S_{2n-1}^2 + 116)^{1/2}) / 2$

4) −1, −7, −8, . . . , $c = 164,$ $S_{2n} = (S_{2n-1} - (5 * S_{2n-1}^2 + 164)^{1/2}) / 2$

5) −1, −8, −9, . . . , $c = 220,$ $S_{2n} = (S_{2n-1} - (5 * S_{2n-1}^2 + 220)^{1/2}) / 2$

6) −1, −9, −10, . . . , $c = 284,$ $S_{2n} = (S_{2n-1} - (5 * S_{2n-1}^2 + 284)^{1/2}) / 2$

EDGAR M. ALEXANDER

7) $-1, -10, -11,$ $c = 356,$ $S_{2n} = (S_{2n-1} - (5 * S_{2n-1}^2 + 356)^{1/2}) / 2$

8) $-1, -11, -12,$ $c = 436,$ $S_{2n} = (S_{2n-1} - (5 * S_{2n-1}^2 + 436)^{1/2}) / 2$

Note that the c value for the negative sequence is the same as for the positive sequence. Also, the formula are similar except the radical is now negative. Therefore, c measures the rate of divergence and not the direction. Interesting enough, the same principles apply to both rational and irrational sequences.

The next table computes the value of c as applied to rational sequences.

1) $1, 0.5, 1.5, 2, \ldots,$ $c = -5,$ $R_{2n} = (R_{2n-1} + (5 * R_{2n-1}^2 - 5)^{1/2}) / 2$

2) $1, 1.5, 2.5, \ldots,$ $c = -1,$ $R_{2n} = (R_{2n-1} + (5 * R_{2n-1}^2 - 1)^{1/2}) / 2$

3) $1, 2.5, 3.5,$ $c = 11,$ $R_{2n} = (R_{2n-1} + (5 * R_{2n-1}^2 + 11)^{1/2}) / 2$

4) $1, 3.5, 4.5, 8,$ $c = 31,$ $R_{2n} = (R_{2n-1} + (5 * R_{2n-1}^2 + 31)^{1/2}) / 2$

5) $1, 4.5, 5.5, 10,$ $c = 59,$ $R_{2n} = (R_{2n-1} + (5 * R_{2n-1}^2 + 59)^{1/2}) / 2$

6) $1, 5.5, 6.5, 12,$ $c = 95,$ $R_{2n} = (R_{2n-1} + (5 * R_{2n-1}^2 + 95)^{1/2}) / 2$

7) $1, 6.5, 7.5,$ $c = 139,$ $R_{2n} = (R_{2n-1} + (5 * R_{2n-1}^2 + 139)^{1/2}) / 2$

This theory is also helpful in creating cross-references between series. Here are two such relationships:

From Fibonacci to Lucas

1) $L_{2n+1} = (5 * F_{2n} + (5 * F_{2n}^2 + 4)^{1/2}) / 2$

2) $L_{2n} = (5 * F_{2n-1} + (5 * F_{2n-1}^2 - 4)^{1/2}) / 2$

From Lucas to Fibonacci

1) $F_{2n+1} = (L_{2n} + ((L_{2n}^2 - 4) / 5)^{1/2}) / 2$

2) $F_{2n} = (L_{2n-1} + ((L_{2n-1}^2 + 4) / 5)^{1/2}) / 2$

There are two special Fibonacci-like sequences where the value of c is zero:

$$c = (2 * a_{n+1} - a_n)^2 - 5 * a_n^2 = 0$$

$$4 * a_{n+1}^2 - 4 * a_{n+1} * a_n + a_n^2 - 5 * a_n^2 = 0$$

$$a_{n+1}^2 - a_{n+1} * a_n - a_n^2 = 0$$

Complete the square for a_{n+1}.

$$a_{n+1}^2 - a_{n+1} * a_n + \tfrac{1}{4} * a_n^2 = 5/4 * a_n^2$$

$$a_{n+1} - \tfrac{1}{2} * a_n = \pm \sqrt{5} / 2 * a_n$$

$$a_{n+1} = (1 \pm \sqrt{5}) / 2 * a_n$$

First Root: $a_{n+1} = (1 + \sqrt{5}) / 2 * a_n = \varphi * a_n$

Fibonacci relationship		*Recursive relationship*
$a_{n+1} = a_n + a_{n-1}$		$a_{n+1} = a_n * \varphi$

. . . .

. . . .

1) 1	$=$	1
2) φ	$=$	φ
3) $1 + \varphi$	$=$	φ^2
4) $1 + 2\varphi$	$=$	φ^3
5) $2 + 3\varphi$	$=$	φ^4
6) $3 + 5\varphi$	$=$	φ^5
7) $5 + 8\varphi$	$=$	φ^6

.

.

.

.

$F(n-1) + F(n) * \varphi$	$=$	φ^n

Note for all $_n$ that the ratio $a_{n+1} / a_n = \varphi$.

EDGAR M. ALEXANDER

Second Root: $a_{n+1} = (1 - \sqrt{5})/2 * a_n = \Phi * a_n$

Fibonacci relationship Recursive relationship $a_{n+1} = a_n + a_{n-1}$ $a_{n+1} = a_n * \Phi$

....
....

1) 1 $=$ 1
2) Φ $=$ Φ
3) $1 + \Phi$ $=$ Φ^2
4) $1 + 2\Phi$ $=$ Φ^3
5) $2 + 3\Phi$ $=$ Φ^4
6) $3 + 5\Phi$ $=$ Φ^5
7) $5 + 8\Phi$ $=$ Φ^6

.

.

$)F(n-1) + F(n) * \Phi = \Phi^n$

Note for all $_n$ **that the ratio** $a_{n+1}/a_n = \Phi$.
This the only special case of Fibonacci-like sequences where $G(0) = 1 * k$,
$G(1) = \Phi * k$, $c = 0$, and the limit of $G(n)/G(n-1) = \Phi$, not φ. Also, this is
the only sequence of its type where the limit of $G(n)$ converges.

....

Calculation of Benet's Formula:

1) $F(n-1) + F(n) * \varphi = \varphi^n$, where $\varphi = (1 + \sqrt{5})/2$
2) $F(n-1) + F(n) * \Phi = \Phi^n$, where $\Phi = (1 - \sqrt{5})/2$
3) $(F(n-1) + F(n) * \varphi) - (F(n-1) + F(n) * \Phi) = \varphi^n - \Phi^n$

4) $F(n) * (\varphi - \Phi) = (\varphi^n - v^n)$

5) $F(n) = (\varphi^n - \Phi^n) / (\varphi - \Phi)$

6) $F(n) = (\varphi^n - \Phi^n) / \sqrt{5}$ Benet's formula

....

....

The Direct Proofs of Four Fibonacci Relationships

We have already shown the following:

$F(n+1) = (F(n) + (5 * F(n)^2 - 4)^{1/2}) / 2$, where n is odd and

$F(n+1) = (F(n)+(5 * F(n)^2 + 4)^{1/2}) / 2$, where n is even.

....

....

Table 1 for n is odd:

a) $F(n)$

b) $F(n+1) = (F(n) + (5 * F(n)^2 - 4)^{1/2}) / 2$

c) $F(n+2) = (3 * F(n) + (5 * F(n)^2 - 4)^{1/2}) / 2$

d) $F(n+3) = (4 * F(n) + 2 * (5 * F(n)^2 - 4)^{1/2}) / 2$

e) $F(n+4) = (7 * F(n) + 3 * (5 * F(n)^2 - 4)^{1/2}) / 2$

f) $F(n+5) = (11 * F(n) + 5 * (5 * F(n)^2 - 4)^{1/2}) / 2$

g) $F(n+6) = (18 * F(n) + 8 * (5 * F(n)^2 - 4)^{1/2}) / 2$

.

.

.

$F(n+k) = (L(k) * F(n) + F(k) * (5 * F(n)^2 - 4)^{1/2}) / 2$

....

Table 2 for *n* is even:

a) $F(n)$

b) $F(n+1) = (F(n) + (5 * F(n)^2 + 4)^{\frac{1}{2}}) / 2$

c) $F(n+2) = (3 * F(n) + (5 * F(n)^2 + 4)^{\frac{1}{2}}) / 2$

d) $F(n+3) = (4 * F(n) + 2 * (5 * F(n)^2 + 4)^{\frac{1}{2}}) / 2$

e) $F(n+4) = (7 * F(n) + 3 * (5 * F(n)^2 + 4)^{\frac{1}{2}}) / 2$

f) $F(n+5) = (11 * F(n) + 5 * (5 * F(n)^2 + 4)^{\frac{1}{2}}) / 2$

g) $F(n+6) = (18 * F(n) + 8 * (5 * F(n)^2 + 4)^{\frac{1}{2}}) / 2$

.

.

.

$$F(n+k) = (L(k) * F(n) + F(k) * (5 * F(n)^2 + 4)^{\frac{1}{2}}) / 2$$

....

Since the direct proof may have a different result depending if *n* is odd or even, both cases have to be proven.

Example 1

For the condition where *n* is odd, prove that $F(n+6) = 4 * F(n+3) + F(n)$

a) $F(n+6) = (18 * F(n) + 8 * (5 * F(n)^2 - 4)^{\frac{1}{2}}) / 2$ "from table 1"

b) $F(n+3) = (4 * F(n) + 2 * (5 * F(n)^2 - 4)^{\frac{1}{2}}) / 2$ "from table 1"

c) $4 * F(n+3) = (16 * F(n) + 8 * (5 * F(n)^2 - 4)^{\frac{1}{2}}) / 2$

d) $4 * F(n+3) + F(n) = (18 * F(n) + 8 * (5 * F(n)^2 - 4)^{\frac{1}{2}}) / 2$

e) item (a) = (d)

Now for the condition where _n_ is even, prove that F(n+6) = 4 * F(n+3) + F(n)

A2) $F(n+6) = (18 * F(n) + 8 * (5 * F(n)^2 + 4)^{1/2}) / 2$ "from table 2"
B2) $F(n+3) = (4 * F(n) + 2 * (5 * F(n)^2 + 4)^{1/2}) / 2$ "from table 2"
C2) $4 * F(n+3) = (16 * F(n) + 8 * (5 * F(n)^2 + 4)^{1/2}) / 2$
D2) $4 * F(n+3) + F(n) = (18 * F(n) + 8 * (5 * F(n)^2 + 4)^{1/2}) / 2$
E2) item (A2) = (D2)

....

Note that this proof consisted of F(n), F(n+3), and F(n+6).

We therefore created the tables using the lowest parameter, F(n). In the next example, F(n–1) is the lowest parameter. So we will use the following tables:

1) If n–1 is odd, then $F(n) = (F(n-1) + (5 * F(n-1)^2 - 4)^{1/2}) / 2$.

The recursive relationship is shown on this table 3.

$F(n-1)$
$F(n) = (F(n-1) + (5 * F(n-1)^2 - 4)^{1/2}) / 2$
$F(n+1) = (3 * F(n-1) + (5 * F(n-1)^2 - 4)^{1/2}) / 2$
$F(n+2) = (4 * F(n-1) + 2 * (5 * F(n-1)^2 - 4)^{1/2}) / 2$
$F(n+3) = (7 * F(n-1) + 3 * (5 * F(n-1)^2 - 4)^{1/2}) / 2$
$F(n+4) = (11 * F(n-1) + 5 * (5 * F(n-1)^2 - 4)^{1/2}) / 2$
$F(n+5) = (18 * F(n-1) + 8 * (5 * F(n-1)^2 - 4)^{1/2}) / 2$

.
.
.

EDGAR M. ALEXANDER

$$F(n-1+k) = (L(k) * F(n-1) + F(k) * (5 * F(n-1)^2 - 4)^{1/2}) / 2$$

. . . .

2) If n–1 is even, then $F(n) = (F(n-1) + (5 * F(n-1)^2 + 4)^{1/2}) / 2$.

The recursive relationship is shown on this table 4.

$F(n-1)$

$$F(n) = (F(n-1) + (5 * F(n-1)^2 + 4)^{1/2}) / 2$$

$$F(n+1) = (3 * F(n-1) + (5 * F(n-1)^2 + 4)^{1/2}) / 2$$

$$F(n+2) = (4 * F(n-1) + 2 * (5 * F(n-1)^2 + 4)^{1/2}) / 2$$

$$F(n+3) = (7 * F(n-1) + 3 * (5 * F(n-1)^2 + 4)^{1/2}) / 2$$

$$F(n+4) = (11 * F(n-1) + 5 * (5 * F(n-1)^2 + 4)^{1/2}) / 2$$

$$F(n+5) = (18 * F(n-1) + 8 * (5 * F(n-1)^2 + 4)^{1/2}) / 2$$

.

.

.

$$F(n-1+k) = (L(k) * F(n-1) + F(k) * (5 * F(n-1)^2 + 4)^{1/2}) / 2$$

Example 2

Where n–1 is odd is the first case

Prove $F(n-1) * F(n+1) - F(n)^2 = (-1)^n$

a) If n–1 odd, then *n* is even
b) $F(n) = (F(n-1) + (5 * F(n-1)^2 - 4)^{1/2}) / 2$ "from table 3"
c) $F(n+1) = (3 * F(n-1) + (5 * F(n-1)^2 - 4)^{1/2}) / 2$ "from table 3"
d) $F(n-1) * F(n+1) = (3 * F(n-1)^2 + F(n-1) * (5 * F(n-1)^2 - 4)^{1/2}) / 2$

e) $F(n)^2 = (F(n-1)^2 + 2 * F(n-1) * (5 * F(n-1)^2 - 4)^{1/2} + 5 * F(n-1)^2 - 4) / 4 = (6 * F(n-1)^2 + 2 * F(n-1) * (5 * F(n-1)^2 - 4)^{1/2} - 4) / 4$

f) $F(n-1) * F(n+1) - F(n)^2 = 4 / 4 = 1 = (-1)^n$

....

Where n–1 is even is the second case

Prove $F(n-1) * F(n+1) - F(n)^2 = (-1)^n$

a) If n–1 even, then *n* is odd.
b) $F(n) = (F(n-1) + (5 * F(n+1)^2 + 4)^{1/2}) / 2$ "from table 4"
c) $F(n+1) = (3 * F(n-1) + (5 * F(n+1)^2 + 4)^{1/2}) / 2$ "from table 4"
d) $F(n-1) * F(n+1) = (3 * F(n-1)^2 + F(n-1) * (5 * F(n-1)^2 + 4)^{1/2}) / 2$
e) $F(n)^2 = (F(n-1)^2 + 2 * F(n-1) * (5 * F(n-1)^2 + 4)^{1/2} + 5 * F(n-1)^2 + 4) / 4 = (6 * F(n-1)^2 + 2 * F(n-1) * (5 * F(n-1)^2 + 4)^{1/2} + 4) / 4$
f) $F(n-1) * F(n+1) - F(n)^2 = -4 / 4 = -1 = (-1)^n$

<u>$F(n-1) * F(n+1) - F(n)^2 = (-1)^n$ holds for both even and odd "n"</u>

<u>*Example 3*</u>

Where n–1 is odd

Prove $L(n)^2 - 5 * F(n)^2 = 4 * (-1)^N$

a) If n–1 is odd, then *n* is even.
b) $L(n) = F(n-1) + F(n+1) = F(n-1) + (3 * F(n-1) + (5 * F(n+1)^2 - 4)^{1/2}) / 2$ "from table 3"

c) $L(n) = (5 * F(n-1) + (5 * F(n+1)^2 - 4)^{1/2}) / 2$

d) $L(n)^2 = (30 * F(n-1)^2 + 10 * (5 * F(n-1)^2 - 4)^{1/2} - 4) / 4$

e) $F(n) = (F(n-1) + (5 * F(n-1)^2 - 4)^{1/2}) / 2$ "from table 3"

f) $F(n)^2 = (6 * F(n-1) + 2 * (F(n-1)^2 - 4)^{1/2} - 4) / 4$ g) $5 * F(n)^2 = (30 * F(n-1) + 10 * (F(n-1)^2 - 4)^{1/2} - 20) / 4$ h) $L(n)^2 - 5 * F(n)^2 = (d) - (g) = (20-4) / 4 = 4 *(-1)^n$

. . . .

Now, where n–1 is even, prove that $L(n)^2 - 5 * F(n)^2 = 4 * (-1)^N$

a) If n–1 is even, then *n* is odd.

b) $L(n) = F(n-1) + F(n+1) = F(n-1) + (3 * F(n-1) + (5 * F(n-1)^2 + 4)^{1/2}) / 2$ "from table 4"

c) $L(n) = (5 * F(n-1) + (5 * F(n-1)^2 + 4)^{1/2}) / 2$

d) $L(n)^2 = (30 * F(n-1)^2 + 10 * (5 * F(n-1)^2 + 4)^{1/2} + 4) / 4$

e) $F(n) = (F(n-1) + (5 * F(n-1)^2 + 4)^{1/2}) / 2$ "from table 4"

f) $F(n)^2 = (6 * F(n-1) + 2 * (F(n-1)^2+4)^{1/2} + 4) / 4$

g) $5 * F(n)^2 = (30 * F(n-1) + 10 * (F(n-1)^2 + 4)^{1/2} + 20) / 4$ h) $L(n)^2 - 5 * F(n)^2 = (d) - (g) = (4 - 20) / 4 = 4 * (-1)^n$

$L(n)^2 - 5 * F(n)^2 = 4 * (-1)^n$ holds for both conditions

Example 4

Where *n* is even or odd, prove $F(2n-1) = F(n-1)^2 + F(n)^2$.

. . . .

Table for *n* even or odd

a) $F(n-1)$

b) $F(n) = (F(n-1) + (5 * F(n-1)^2 + c)^{1/2}) / 2$

c) $F(n+1) = (3 * F(n-1) + (5 * F(n-1)^2 + c)^{1/2}) / 2$

d) $F(n+2) = (4 * F(n-1) + 2 * (5 * F(n-1)^2 + c)^{1/2}) / 2$

e) $F(n+3) = (7 * F(n-1) + 3 * (5 * F(n-1)^2 + c)^{1/2}) / 2$

f) $F(n+4) = (11 * F(n-1) + 5 * (5 * F(n-1)^2 + c)^{1/2}) / 2$

.

.

. $F(n-1+k) = (L(k) * F(n-1) + F(k) * (5 * F(n-1)^2 + c)^{1/2}) / 2$

....

a) $F(2n-1) = F(n-1 + n) =$
   $(L(n) * F(n-1) + F(n) * (5 * F(n)^2 + c)^{1/2}) / 2$

b) $L(n) * F(n-1) = (F(n-1) + F(n+1)) * F(n-1)$

c) $L(n) = F(n+1) + F(n-1)$ "definition"

d) $F(n+1) + F(n-1) = (5 * F(n-1) + (5 * F(n-1)^2 + c)^{1/2}) / 2$ "from table 3 and 4"

e) $(L(n) * F(n-1)) / 2 = (5 * F(n-1)^2 + F(n-1) * (5 * F(n-1)^2 + c)^{1/2}) / 4$

f) $(F(n) * (5 * F(n-1)^2 + c)^{1/2}) / 2 =$
   $(F(n-1) * (5 * F(n-1)^2 + c)^{1/2} + 5 * F(n-1)^2 + c) / 4$

g) $(e) + (f) = F(2n-1) =$
   $(10 * F(n-1)^2 + 2 * F(n-1) * (5 * F(n-1)^2 + c)^{1/2} + c) / 4$

h) $F(n)^2 = (6 * F(n-1)^2 + 2 * F(n-1) * (5 * F(n-1)^2 + c)^{1/2} + c) / 4$

i) $F(n-1)^2 + F(n)^2 = (10 * F(n-1)^2 + 2 * F(n-1) * (5 * F(n-1)^2 + c)^{1/2} + c) / 4$

j) $(g) = (i)$ for both cases of $c = -4$ and $c = 4$.

k) $F(2n-1) = F(n-1)^2 + F(n)^2$

....

EDGAR M. ALEXANDER

The following tables are provided to be used in other proofs that you may have:

1. F(n–6) where n–6 is odd

a) $F(n–6)$
b) $F(n–5) = (F(n–6) + (5 * F(n–6)^2 – 4)^{1/2}) / 2$
c) $F(n–4) = (3 * F(n–6) + (5 * F(n–6)^2 – 4)^{1/2}) / 2$
d) $F(n–3) = (4 * F(n–6) + 2 * (5 * F(n–6)^2 – 4)^{1/2}) / 2$
e) $F(n–2) = (7 * F(n–6) + 3 * (5 * F(n–6)^2 – 4)^{1/2}) / 2$
f) $F(n–1) = (11 * F(n–6) + 5 * (5 * F(n–6)^2 – 4)^{1/2}) / 2$
g) $F(n) = (18 * F(n–6) + 8 * (5 * F(n–6)^2 – 4)^{1/2}) / 2$
h) $F(n+1) = (29 * F(n–6) + 13 * (5 * F(n–6)^2 – 4)^{1/2}) / 2$
i) $F(n+2) = (47 * F(n–6) + 21 * (5 * F(n–6)^2 – 4)^{1/2}) / 2$
j) $F(n+3) = (76 * F(n–6) + 34 * (5 * F(n–6)^2 – 4)^{1/2}) / 2$
k) $F(n+4) = (123 * F(n–6) + 55 * (5 * F(n–6)^2 – 4)^{1/2}) / 2$
l) $F(n+5) = (199 * F(n–6) + 89 * (5 * F(n–6)^2 – 4)^{1/2}) / 2$
m) $F(n+6) = (322 * F(n–6) + 144 * (5 * F(n–6)^2 – 4)^{1/2}) / 2$

....

....

1. F(n–5) where n–5 is odd

a) $F(n–5)$
b) $F(n–4) = (F(n–5) + (5 * F(n–5)^2 – 4)^{1/2}) / 2$
c) $F(n–3) = (3 * F(n–5) + (5 * F(n–5)^2 – 4)^{1/2}) / 2$
d) $F(n–2) = (4 * F(n–5) + 2 * (5 * F(n–5)^2 – 4)^{1/2}) / 2$
e) $F(n–1) = (7 * F(n–5) + 3 * (5 * F(n–5)^2 – 4)^{1/2}) / 2$
f) $F(n) = (11 * F(n–5) + 5 * (5 * F(n–5)^2 – 4)^{1/2}) / 2$

g) $F(n+1) = (18 * F(n-5) + 8 * (5 * F(n-5)^2 - 4)^{\frac{1}{2}}) / 2$

h) $F(n+2) = (29 * F(n-5) + 13 * (5 * F(n-5)^2 - 4)^{\frac{1}{2}}) / 2$

i) $F(n+3) = (47 * F(n-5) + 21 * (5 * F(n-5)^2 - 4)^{\frac{1}{2}}) / 2$

j) $F(n+4) = (76 * F(n-5) + 34 * (5 * F(n-5)^2 - 4)^{\frac{1}{2}}) / 2$

k) $F(n+5) = (123 * F(n-5) + 55 * (5 * F(n-5)^2 - 4)^{\frac{1}{2}}) / 2$

l) $F(n+6) = (199 * F(n-5) + 89 * (5 * F(n-5)^2 - 4)^{\frac{1}{2}}) / 2$

m) $F(n+7) = (322 * F(n-5) + 144 * (5 * F(n-5)^2 - 4)^{\frac{1}{2}}) / 2$

....

....

1. F(n–4) where n–4 is odd

a) $F(n-4)$

b) $F(n-3) = (F(n-4) + (5 * F(n-4)^2 - 4)^{\frac{1}{2}}) / 2$

c) $F(n-2) = (3 * F(n-4) + (5 * F(n-4)^2 - 4)^{\frac{1}{2}}) / 2$

d) $F(n-1) = (4 * F(n-4) + 2 * (5 * F(n-4)^2 - 4)^{\frac{1}{2}}) / 2$

e) $F(n) = (7 * F(n-4) + 3 * (5 * F(n-4)^2 - 4)^{\frac{1}{2}}) / 2$

f) $F(n+1) = (11 * F(n-4) + 5 * (5 * F(n-4)^2 - 4)^{\frac{1}{2}}) / 2$

g) $F(n+2) = (18 * F(n-4) + 8 * (5 * F(n-4)^2 - 4)^{\frac{1}{2}}) / 2$

h) $F(n+3) = (29 * F(n-4) + 13 * (5 * F(n-4)^2 - 4)^{\frac{1}{2}}) / 2$

i) $F(n+4) = (47 * F(n-4) + 21 * (5 * F(n-4)^2 - 4)^{\frac{1}{2}}) / 2$

j) $F(n+5) = (76 * F(n-4) + 34 * (5 * F(n-4)^2 - 4)^{\frac{1}{2}}) / 2$

k) $F(n+6) = (123 * F(n-4) + 55 * (5 * F(n-4)^2 - 4)^{\frac{1}{2}}) / 2$

l) $F(n+7) = (199 * F(n-4) + 89 * (5 * F(n-4)^2 - 4)^{\frac{1}{2}}) / 2$

m) $F(n+8) = (322 * F(n-4) + 144 * (5 * F(n-4)^2 - 4)^{\frac{1}{2}}) / 2$

....

....

EDGAR M. ALEXANDER

1. F(n–3) where n–3 is odd

a) $F(n-3)$

b) $F(n-2) = (F(n-3) + (5 * F(n-3)^2 - 4)^{1/2}) / 2$

c) $F(n-1) = (3 * F(n-3) + (5 * F(n-3)^2 - 4)^{1/2}) / 2$

d) $F(n) = (4 * F(n-3) + 2 * (5 * F(n-3)^2 - 4)^{1/2}) / 2$

e) $F(n+1) = (7 * F(n-3) + 3 * (5 * F(n-3)^2 - 4)^{1/2}) / 2$

f) $F(n+2) = (11 * F(n-3) + 5 * (5 * F(n-3)^2 - 4)^{1/2}) / 2$

g) $F(n+3) = (18 * F(n-3) + 8 * (5 * F(n-3)^2 - 4)^{1/2}) / 2$

h) $F(n+4) = (29 * F(n-3) + 13 * (5 * F(n-3)^2 - 4)^{1/2}) / 2$

i) $F(n+5) = (47 * F(n-3) + 21 * (5 * F(n-3)^2 - 4)^{1/2}) / 2$

j) $F(n+6) = (76 * F(n-3) + 34 * (5 * F(n-3)^2 - 4)^{1/2}) / 2$

k) $F(n+7) = (123 * F(n-3) + 55 * (5 * F(n-3)^2 - 4)^{1/2}) / 2$

l) $F(n+8) = (199 * F(n-3) + 89 * (5 * F(n-3)^2 - 4)^{1/2}) / 2$

m) $F(n+9) = (322 * F(n-3) + 144 * (5 * F(n-3)^2 - 4)^{1/2}) / 2$

....

....

1. F(n–2) where n–2 is odd

a) $F(n-2)$

b) $F(n-1) = (F(n-2) + (5 * F(n-2)^2 - 4)^{1/2}) / 2$

c) $F(n) = (3 * F(n-2) + (5 * F(n-2)^2 - 4)^{1/2}) / 2$

d) $F(n+1) = (4 * F(n-2) + 2 * (5 * F(n-2)^2 - 4)^{1/2}) / 2$

e) $F(n+2) = (7 * F(n-2) + 3 * (5 * F(n-2)^2 - 4)^{1/2}) / 2$

f) $F(n+3) = (11 * F(n-2) + 5 * (5 * F(n-2)^2 - 4)^{1/2}) / 2$

g) $F(n+4) = (18 * F(n-2) + 8 * (5 * F(n-2)^2 - 4)^{1/2}) / 2$

h) $F(n+5) = (29 * F(n-2) + 13 * (5 * F(n-2)^2 - 4)^{1/2}) / 2$

i) $F(n+6) = (47 * F(n-2) + 21 * (5 * F(n-2)^2 - 4)^{1/2}) / 2$

j) $F(n+7) = (76 * F(n-2) + 34 * (5 * F(n-2)^2 - 4)^{1/2}) / 2$

k) $F(n+8) = (123 * F(n-2) + 55 * (5 * F(n-2)^2 - 4)^{1/2}) / 2$

l) $F(n+9) = (199 * F(n-2) + 89 * (5 * F(n-2)^2 - 4)^{1/2}) / 2$

m) $F(n+10) = (322 * F(n-2) + 144 * (5 * F(n-2)^2 - 4)^{1/2}) / 2$

....

....

1. F(n–1) where n–1 is odd

a) $F(n-1)$

b) $F(n) = (F(n-1) + (5 * F(n-1)^2 - 4)^{1/2}) / 2$

c) $F(n+1) = (3 * F(n-1) + (5 * F(n-1)^2 - 4)^{1/2}) / 2$

d) $F(n+2) = (4 * F(n-1) + 2 * (5 * F(n-1)^2 - 4)^{1/2}) / 2$

e) $F(n+3) = (7 * F(n-1) + 3 * (5 * F(n-1)^2 - 4)^{1/2}) / 2$

f) $F(n+4) = (11 * F(n-1) + 5 * (5 * F(n-1)^2 - 4)^{1/2}) / 2$

g) $F(n+5) = (18 * F(n-1) + 8 * (5 * F(n-1)^2 - 4)^{1/2}) / 2$

h) $F(n+6) = (29 * F(n-1) + 13 * (5 * F(n-1)^2 - 4)^{1/2}) / 2$

i) $F(n+7) = (47 * F(n-1) + 21 * (5 * F(n-1)^2 - 4)^{1/2}) / 2$

j) $F(n+8) = (76 * F(n-1) + 34 * (5 * F(n-1)^2 - 4)^{1/2}) / 2$

k) $F(n+9) = (123 * F(n-1) + 55 * (5 * F(n-1)^2 - 4)^{1/2}) / 2$

l) $F(n+10) = (199 * F(n-1) + 89 * (5 * F(n-1)^2 - 4)^{1/2}) / 2$

m) $F(n+11) = (322 * F(n-1) + 144 * (5 * F(n-1)^2 - 4)^{1/2}) / 2$

....

....

1. F(n) where n is odd

a) $F(n)$

b) $F(n+1) = (F(n) + (5 * F(n)^2 - 4)^{1/2}) / 2$

c) $F(n+2) = (3 * F(n) + (5 * F(n)^2 - 4)^{1/2}) / 2$

d) $F(n+3) = (4 * F(n) + 2 * (5 * F(n)^2 - 4)^{1/2}) / 2$

e) $F(n+4) = (7 * F(n) + 3 * (5 * F(n)^2 - 4)^{1/2}) / 2$

f) $F(n+5) = (11 * F(n) + 5 * (5 * F(n)^2 - 4)^{1/2}) / 2$

g) $F(n+6) = (18 * F(n) + 8 * (5 * F(n)^2 - 4)^{1/2}) / 2$

h) $F(n+7) = (29 * F(n) + 13 * (5 * F(n)^2 - 4)^{1/2}) / 2$

i) $F(n+8) = (47 * F(n) + 21 * (5 * F(n)^2 - 4)^{1/2}) / 2$

j) $F(n+9) = (76 * F(n) + 34 * (5 * F(n)^2 - 4)^{1/2}) / 2$

k) $F(n+10) = (123 * F(n) + 55 * (5 * F(n)^2 - 4)^{1/2}) / 2$

l) $F(n+11) = (199 * F(n) + 89 * (5 * F(n)^2 - 4)^{1/2}) / 2$

m) $F(n+12) = (322 * F(n) + 144 * (5 * F(n)^2 - 4)^{1/2}) / 2$

....

The following tables are provided to be used where the leading term is even:

1. F(n–6) where n–6 is even

a) $F(n-6)$

b) $F(n-5) = (F(n-6) + (5 * F(n-6)^2 + 4)^{1/2}) / 2$

c) $F(n-4) = (3 * F(n-6) + (5 * F(n-6)^2 + 4)^{1/2}) / 2$

d) $F(n-3) = (4 * F(n-6) + 2 * (5 * F(n-6)^2 + 4)^{1/2}) / 2$

e) $F(n-2) = (7 * F(n-6) + 3 * (5 * F(n-6)^2 + 4)^{1/2}) / 2$

f) $F(n-1) = (11 * F(n-6) + 5 * (5 * F(n-6)^2 + 4)^{1/2}) / 2$

g) $F(n) = (18 * F(n-6) + 8 * (5 * F(n-6)^2 + 4)^{1/2}) / 2$

h) $F(n+1) = (29 * F(n-6) + 13 * (5 * F(n-6)^2 + 4)^{1/2}) / 2$

i) $F(n+2) = (47 * F(n-6) + 21 * (5 * F(n-6)^2 + 4)^{1/2}) / 2$

j) $F(n+3) = (76 * F(n-6) + 34 * (5 * F(n-6)^2 + 4)^{1/2}) / 2$

k) $F(n+4) = (123 * F(n-6) + 55 * (5 * F(n-6)^2 + 4)^{1/2}) / 2$

l) $F(n+5) = (199 * F(n-6) + 89 * (5 * F(n-6)^2 + 4)^{1/2}) / 2$

m) $F(n+6) = (322 * F(n-6) + 144 * (5 * F(n-6)^2 + 4)^{1/2}) / 2$

....

....

1. F(n–5) where n–5 is even

a) $F(n-5)$

b) $F(n-4) = (F(n-5) + (5 * F(n-5)^2 + 4)^{1/2}) / 2$

c) $F(n-3) = (3 * F(n-5) + (5 * F(n-5)^2 + 4)^{1/2}) / 2$

d) $F(n-2) = (4 * F(n-5) + 2 * (5 * F(n-5)^2 + 4)^{1/2}) / 2$

e) $F(n-1) = (7 * F(n-5) + 3 * (5 * F(n-5)^2 + 4)^{1/2}) / 2$

f) $F(n) = (11 * F(n-5) + 5 * (5 * F(n-5)^2 + 4)^{1/2}) / 2$

g) $F(n+1) = (18 * F(n-5) + 8 * (5 * F(n-5)^2 + 4)^{1/2}) / 2$

h) $F(n+2) = (29 * F(n-5) + 13 * (5 * F(n-5)^2 + 4)^{1/2}) / 2$

i) $F(n+3) = (47 * F(n-5) + 21 * (5 * F(n-5)^2 + 4)^{1/2}) / 2$

j) $F(n+4) = (76 * F(n-5) + 34 * (5 * F(n-5)^2 + 4)^{1/2}) / 2$

k) $F(n+5) = (123 * F(n-5) + 55 * (5 * F(n-5)^2 + 4)^{1/2}) / 2$

l) $F(n+6) = (199 * F(n-5) + 89 * (5 * F(n-5)^2 + 4)^{1/2}) / 2$

m) $F(n+7) = (322 * F(n-5) + 144 * (5 * F(n-5)^2 + 4)^{1/2}) / 2$

....

....

EDGAR M. ALEXANDER

1. F(n–4) where n–4 is even

a) $F(n–4)$
b) $F(n–3) = (F(n–4)(5 * F(n–4)^2 + 4)^{1/2}) / 2$
c) $F(n–2) = (3 * F(n–4)(5 * F(n–4)^2 + 4)^{1/2}) / 2$
d) $F(n–1) = (4 * F(n–4)2 * (5 * F(n–4)^2 + 4)^{1/2}) / 2$
e) $F(n) = (7 * F(n–4)3 * (5 * F(n–4)^2 – 4)^{1/2}) / 2$
f) $F(n+1) = (11 * F(n–4)5 * (5 * F(n–4)^2 + 4)^{1/2}) / 2$
g) $F(n+2) = (18 * F(n–4)8 * (5 * F(n–4)^2 + 4)^{1/2}) / 2$
h) $F(n+3) = (29 * F(n–4)13 * (5 * F(n–4)^2 + 4)^{1/2}) / 2$
i) $F(n+4) = (47 * F(n–4)21 * (5 * F(n–4)^2 + 4)^{1/2}) / 2$
j) $F(n+5) = (76 * F(n–4)34 * (5 * F(n–4)^2 + 4)^{1/2}) / 2$
k) $F(n+6) = (123 * F(n–4)55 * (5 * F(n–4)^2 + 4)^{1/2}) / 2$
l) $F(n+7) = (199 * F(n–4)89 * (5 * F(n–4)^2 + 4)^{1/2}) / 2$ l) $F(n+8) = (322 * F(n–4)144 * (5 * F(n–4)^2 + 4)^{1/2}) / 2$

....

....

1. F(n–3) where n–3 is even

a) $F(n–3)$
b) $F(n–2) = (F(n–3) + (5 * F(n–3)^2 + 4)^{1/2}) / 2$
c) $F(n–1) = (3 * F(n–3) + (5 * F(n–3)^2 + 4)^{1/2}) / 2$
d) $F(n) = (4 * F(n–3) + 2 * (5 * F(n–3)^2 + 4)^{1/2}) / 2$
e) $F(n+1) = (7 * F(n–3) + 3 * (5 * F(n–3)^2 + 4)^{1/2}) / 2$
f) $F(n+2) = (11 * F(n–3) + 5 * (5 * F(n–3)^2 + 4)^{1/2}) / 2$
g) $F(n+3) = (18 * F(n–3) + 8 * (5 * F(n–3)^2 + 4)^{1/2}) / 2$
h) $F(n+4) = (29 * F(n–3) + 13 * (5 * F(n–3)^2 + 4)^{1/2}) / 2$

i) $F(n+5) = (47 * F(n-3) + 21 * (5 * F(n-3)^2 + 4)^{1/2}) / 2$

j) $F(n+6) = (76 * F(n-3) + 34 * (5 * F(n-3)^2 + 4)^{1/2}) / 2$

k) $F(n+7) = (123 * F(n-3) + 55 * (5 * F(n-3)^2 + 4)^{1/2}) / 2$

l) $F(n+8) = (199 * F(n-3) + 89 * (5 * F(n-3)^2 + 4)^{1/2}) / 2$ l) $F(n+9) = (322 * F(n-3) + 144 * (5 * F(n-3)^2 + 4)^{1/2}) / 2$

....
....

1. F(n–2) where n–2 is even

a) $F(n-2)$

b) $F(n-1) = (F(n-2) + (5 * F(n-2)^2 + 4)^{1/2}) / 2$

c) $F(n) = (3 * F(n-2) + (5 * F(n-2)^2 + 4)^{1/2}) / 2$

d) $F(n+1) = (4 * F(n-2) + 2 * (5 * F(n-2)^2 + 4)^{1/2}) / 2$

e) $F(n+2) = (7 * F(n-2) + 3 * (5 * F(n-2)^2 + 4)^{1/2}) / 2$

f) $F(n+3) = (11 * F(n-2) + 5 * (5 * F(n-2)^2 + 4)^{1/2}) / 2$

g) $F(n+4) = (18 * F(n-2) + 8 * (5 * F(n-2)^2 + 4)^{1/2}) / 2$

h) $F(n+5) = (29 * F(n-2) + 13 * (5 * F(n-2)^2 + 4)^{1/2}) / 2$

i) $F(n+6) = (47 * F(n-2) + 21 * (5 * F(n-2)^2 + 4)^{1/2}) / 2$

j) $F(n+7) = (76 * F(n-2) + 34 * (5 * F(n-2)^2 + 4)^{1/2}) / 2$

k) $F(n+8) = (123 * F(n-2) + 55 * (5 * F(n-2)^2 + 4)^{1/2}) / 2$

l) $F(n+9) = (199 * F(n-2) + 89 * (5 * F(n-2)^2 + 4)^{1/2}) / 2$

m) $F(n+10) = (322 * F(n-2) + 144 * (5 * F(n-2)^2 + 4)^{1/2}) / 2$

....
....

EDGAR M. ALEXANDER

1. F(n1) where n–1 is even

a) $F(n-1)$

b) $F(n) = (F(n-1) + (5 * F(n-1)^2 + 4)^{1/2}) / 2$

c) $F(n+1) = (3 * F(n-1) + (5 * F(n-1)^2 + 4)^{1/2}) / 2$

d) $F(n+2) = (4 * F(n-1) + 2 * (5 * F(n-1)^2 + 4)^{1/2}) / 2$

e) $F(n+3) = (7 * F(n-1) + 3 * (5 * F(n-1)^2 + 4)^{1/2}) / 2$

f) $F(n+4) = (11 * F(n-1) + 5 * (5 * F(n-1)^2 + 4)^{1/2}) / 2$

g) $F(n+5) = (18 * F(n-1) + 8 * (5 * F(n-1)^2 + 4)^{1/2}) / 2$

h) $F(n+6) = (29 * F(n-1) + 13 * (5 * F(n-1)^2 + 4)^{1/2}) / 2$

i) $F(n+7) = (47 * F(n-1) + 21 * (5 * F(n-1)^2 + 4)^{1/2}) / 2$

j) $F(n+8) = (76 * F(n-1) + 34 * (5 * F(n-1)^2 + 4)^{1/2}) / 2$

k) $F(n+9) = (123 * F(n-1) + 55 * (5 * F(n-1)^2 + 4)^{1/2}) / 2$

l) $F(n+10) = (199 * F(n-1) + 89 * (5 * F(n-1)^2 + 4)^{1/2}) / 2$

m) $F(n+11) = (322 * F(n-1) + 144 * (5 * F(n-1)^2 + 4)^{1/2}) / 2$

....

....

1. F(n) where n is even

a) $F(n)$

b) $F(n+1) = (F(n) + (5 * F(n)^2 + 4)^{1/2}) / 2$

c) $F(n+2) = (3 * F(n) + (5 * F(n)^2 + 4)^{1/2}) / 2$

d) $F(n+3) = (4 * F(n) + 2 * (5 * F(n)^2 + 4)^{1/2}) / 2$

e) $F(n+4) = (7 * F(n) + 3 * (5 * F(n)^2 + 4)^{1/2}) / 2$

f) $F(n+5) = (11 * F(n) + 5 * (5 * F(n)^2 + 4)^{1/2}) / 2$

g) $F(n+6) = (18 * F(n) + 8 * (5 * F(n)^2 + 4)^{1/2}) / 2$

h) $F(n+7) = (29 * F(n) + 13 * (5 * F(n)^2 + 4)^{1/2}) / 2$

i) $F(n+8) = (47 * F(n) + 21 * (5 * F(n)^2 + 4)^{1/2}) / 2$

j) $F(n+9) = (76 * F(n) + 34 * (5 * F(n)^2 + 4)^{\frac{1}{2}}) / 2$

k) $F(n+10) = (123 * F(n) + 55 * (5 * F(n)^2 + 4)^{\frac{1}{2}}) / 2$

l) $F(n+11) = (199 * F(n) + 89 * (5 * F(n)^2 + 4)^{\frac{1}{2}}) / 2$ l) $F(n+12) = (322 * F(n) + 144 * (5 * F(n)^2 + 4)^{\frac{1}{2}}) / 2$

EDGAR M. ALEXANDER

Exercises

We have discussed and developed many of the tools necessary for solving direct proofs. In this section we will discover the thinking behind the creation of some of the formula that were used. They seemed to have come from nowhere and appear. They didn't just happen. And here you'll be asked to step through some of the creations.

This is how the creation of "F(n+1)" was developed from "F(n)". Also it is where Geometry and Number Theory cross paths. In the book "In Search of PI" the following arctan relationship is proven:

Theorem 4 *(from "In Search of PI")*

If 'a' is any positive integer then there exists integers 'x' and 'u' such that:

arcTan(1/a)=arc Tan(1/u) + arc Tan(1/(x+a));

$u = (a^2 +ax+1) / x = (a^2 + 1) / x + a$; *where 'x' is a integer factor of '$a^2 + 1$'*

This theorem shows the rule for expanding any arctan. Note that only integer factors of ($a^2 + 1$) can be used to compute u.

Example 1.

1) a = 2
2) *This implies that* u = 5/x + 2
3) *Positive factors of '5' are:* '1' *and* '5'.
4) Let 'x =1'
5) u = 7

6) 'a + x = 3'

7) arc Tan(1/2) = arc Tan(1/3) + arc Tan(1/7)

8) Let 'x = 5'

9) 'u = 1+2 = 3'

10) 'a + x = 2 + 5 = 7'

11) The only two angle arctan formula of
'arc Tan(1/2)' is

arc Tan(1/2) = arc Tan(1/3) + arc Tan(1/7)

Note that 'x = positive factors of $(a^2 + 1)$' yield forward expansion of 'arc Tan(1/2'.

....
....

12) *Negative factors of '5' are: '-1' and '-5'.*

13) Let 'x = -1'

14) u = -5 + 2 = -3

15) 'a + x = 2-1 = 1'

16) arc Tan(1/2) = arc Tan(-1/3) + arc Tan(1)

17) arc Tan(1/2) = -arc Tan(1/3) + arc Tan(1)

18) arc Tan(1/2) + arc Tan(1/3) = arc Tan(1)

19) Let 'x = -5'

20) u = -1 + 2 = 1

21) 'a + x = 2 – 5 = -3'

22) arc Tan(1/2) = arc Tan(1) + arc Tan(-1/3)

23) arc Tan(1/2) = arc Tan(1) -arc Tan(1/3)

24) arc Tan(1/2) + arc Tan(1/3) = arc Tan(1)

EDGAR M. ALEXANDER

By using negative factors of $(a^2 + 1)$, arc $\text{Tan}(1/2)$ is not expanded. But it appears as a portion of the arc $\text{Tan}(1)$ expansion.

The above proofs are related to Geometry relationships. At the same time Fibonacci knew of the arctan relationship:

For any even Fibonacci number, '$F(2n)$', *arc $Tan(1/F(2n))$ = arc Tan(1/F(2n+1)) + arc Tan(1/F(2n+2))*. At this point I will developed **Theorem 5** for consecutive Fibonacci Numbers: where

1) 'a', 'u' and '$x + a$' are consecutive Fibonacci Numbers.
2) 'a' is an even Fibonacci Number &

   : $a + u = x + a$, $u = x$

   from definition of consecutive Fibonacci Numbers.
3) $x = (a^2 + 1) / x + a$
4) $x^2 = ax + 1 + a^2$; *complete the square for x*
5) $x^2 - ax + a^2/4 = a^2/4 + 1 + a^2$
6) $x = ((5 a^2 + 4)/4)^{1/2} + a/2$

Theorem 5 *(from "In Search of PI")*

If 'a' is an even Fibonacci Number then:

a, x; $(a + x)$ are consecutive Fibonacci where $x = ((5 a^2 + 4) / 4)^{1/2} + a/2$.

example 1. a = 3 then x $= \underline{((5 * 3^2 + 4) / 4)^{1/2} + 3/2 = 5}$; $\underline{\& (a + x) = (3 + 5) = 8}$

Corollary 5.1 *(created in "In Search of PI")*

1) $x^2 = ax + 1 + a^2$ was used in **Theorem 5**.

In *this time complete the square for a*

If 'x' is an odd Fibonacci Number then: a, x; (a + x) are consecutive Fibonacci where a = $((5\ x^2-4)/4)^{1/2}$-x /2.

example 1. x = 5 then a = $((5*\ 5^2-4)/4)^{1/2}$-5 /2 = 3 & (a + x) = (3 + 5) = 8

This concludes how the creation of "F(n+1)" evolved from "F(n)". It used an arctan relationship to develop a Fibonacci Relationship. However the flow of information for developing proofs go in both direction. We will now develop an arctan relationship from Fibonacci Proofs.

A famous mathematician, Lucas, proved the *Following 'two arc Tan expansion' for odd Fibonacci Numbers:*

arc Tan(1 / F(2n+1)) = arc Tan(1/L(2N)) + arc Tan(1 / L(2N+2)).

This relationship can be developed or proven as follows:

1) If 'a' is any positive integer then there exists integers 'x' and 'u' such that: arcTan(1/a)=arcTan(1/(x+a))+arcTan(1/u);u=$(a^2+ax+1)/x=(a^2+1)/x+a$; *where 'x' is a integer factor of '$a^2 + 1$'*. . . . see from **Theorem 4**
2) Let 'a' = F(2N + 1)
3) $F^2(2n+1)$ + 1= F(2n-1) * F(2n+3) is a known relationship.
4) U = $(a^2 + 1)$ / x + a = (F(2n-1) * F(2n+3) / x) + F(2N + 1)
5) 'x + a' = x + F(2N + 1)
6) if we let x = F(2n-1) then 'U = F(2n+3) + F(2N + 1)' & 'x + a' = F(2n-1) + F(2N + 1)
7) F(2n-1) + F(2N + 1) = L(2N); F(2n+1) + F(2N + 3) = L(2N+2)

EDGAR M. ALEXANDER

8) This leads to the conclusion: arc $\text{Tan}(1 / F(2N + 1)) = \text{arc Tan}(1 / L(2N)) + \text{arc Tan}(1 / L(2N+2)) \ldots$ *Lucas formula*

The same result would be reached by setting '$x = F(2n+3)$'. This is one of the most famous 'two arc Tan' expansions.

On the following pages are a list of exercises that should be completed by the reader. Each exercise will consist of an existing formula, hints of how to prove the formula and any sub formulation needed. The reader is would then use the trailing pages to develop the proof.

Exercise 1

1) You are given that this formula is already proven: $F^2(2N) = F(2N-1)*F(2N+1)$; and $\arctan(1/a) = \arctan(1/(x+a)) + \arctan(1/u)$;

$u = (a^2 + ax + 1) / x = (a^2 + 1) / x + a$ *where 'x' is a integer factor of 'a2 + 1' from* **Theorem 4**

2) Fibonacci's formula is: For any even Fibonacci number, '$F(2n)$', *arc Tan(1/F(2n)) = arc Tan(1/F(2n+1)) + arc Tan(1/F(2n+2)).*

3) Use '$a = F(2N)$' and prove Fibonacci's formula.

Exercise 1 (continued)

Exercise 2

1) In this exercise you are given 'a' is an even Fibonacci Number where:
$F(2n) = $ 'a' and $F(2n+1) = (5 a^2 + 4)^{1/2}/2 + a /2$.

2) Prove that the limit of '$F(2N+k)/F(2N)$' as n goes to infinity is '$(L(k)+F(k)*5^{.5})/2$.

3) Prove the same conditions hold for '$F(2n+1)$'

Exercise 2 (continued)

Exercise 3

1) Use the fact that the limit of '$(F(2n+k+1)/F(2n+k)) = \varphi$' & the limit of '$(F(2n + k + 1)/F(2n + k)) = (L(k+1)+F(k+1)*5^{.5})/(L(k)+F(k)*5^{.5})$.

2) Prove that $(L(n)+F(n)*5^{.5})/2 = \varphi^n$

Exercise 3 (continued)

Exercise 4

1) You are given that $(1-\varphi) = 1 - ((L(k)+F(k)*5^{.5}) / (L(k-1)+F(k-1)*5^{.5}))$.

2) Show that $(1-\varphi) = -1*((L(n-1)+F(n-1)*5^{.5}) / (L(n)+F(n)*5^{.5}))$.

EDGAR M. ALEXANDER

Exercise 4 (continued)

Exercise 5

1) You are given the results of *exercise 4*.
Prove that $(1-\varphi)^n = (-1)^n * (2/ (L(n)+F(n)*5^{.5}))$.

Exercise 5 (continued)

Exercise 6

1) You are given that $L(n)^2 - F(n)^2 * 5 = ((-1)^n * 4)$ and $(1-\varphi)^n = (-1)^n * (2/(L(n)+F(n)*5^{.5}))$.

2) Prove that $-(1-\varphi)^n = (F(n)*5^{.5} - L(n))/2$

Exercise 7

1) You are given that $(L(n)+F(n)*5^{.5}) / 2 = \varphi^n$ and $-(1-\varphi)^n = (F(n)*5^{.5}-L(n))/2$.

2) Prove Benet's equation.

Exercise 7 (continued)

1) Prove that $L(n) = \varphi^n + (1-\varphi)^n$

EDGAR M. ALEXANDER

SUMMARY

THIS BOOK IS directed to the audience of high school and college students who are interested in numbers, math sequences, or Fibonacci numbers generally. After reading this paper, you will be better prepared about the subject.

Although the direct proof is not intended to replace proof by inductive reasoning, it, in many cases, does offer a helpful alternative. Formulas are reduced to solving one equation with one unknown.